THE PLANETS OF OUR SOLAR SYSTEM

NEPTUNE

by Meg Thacher

BrightPoint Press

San Diego, CA

BrightPoint Press

an imprint of ReferencePoint Press, Inc.
Printed in the United States

For more information, contact:
BrightPoint Press
PO Box 27779
San Diego, CA 92198
www.BrightPointPress.com

LIBRARY OF CONGRESS CATALOGING-IN-PUBLICATION DATA

Name: Thacher, Meg, author.
Title: Neptune / by Meg Thacher.
Description: San Diego, CA: ReferencePoint Press, 2026 | Series: The planets of our solar system | Audience: Grade 7 to 9 | Includes bibliographical references and index.
Identifiers: ISBN: 9781678211684 (hardcover) | ISBN: 9781678211691 (eBook)
The complete Library of Congress record is available at www.loc.gov.

CONTENTS

AT A GLANCE

- Neptune is the solar system's most distant planet from the Sun. It is thirty times farther from the Sun than Earth on average.

- Neptune and Uranus are types of planets known as ice giants. They belong to a larger planet family called giant planets.

- Neptune is four times larger in diameter than Earth. Its rocky core is surrounded by hydrogen-rich gases.

- Neptune has giant storms and the fastest winds in the solar system.

- Scientists discovered Neptune in 1846 after predicting its existence the previous year. The prediction was based on how the undiscovered planet's gravity was affecting the other planets' motion.

- Neptune is named after the Roman god of the sea. Its moons are named after minor Greek sea gods.

- *Voyager 2* is the only space probe to have visited Neptune. It discovered a great deal about Neptune during its short visit.

- A group of moons called regular moons formed along with Neptune. The planet's irregular moons are captured asteroids and comets.

- Neptune's largest moon is Triton. It was probably once a dwarf planet.

THE END OF A LONG VOYAGE

People all over Earth sat on the edges of their seats. They watched as picture after picture appeared on their TVs. It was the night of August 24, 1989. The *Voyager 2* spacecraft was sending its first pictures of the planet Neptune back to Earth.

Astronomers at the National Aeronautics and Space Administration (NASA) stayed up all night. They continued to release images and data for the next

A *Titan IIIE-Centaur* rocket took *Voyager 2* into space on August 20, 1977.

UNITED STATES

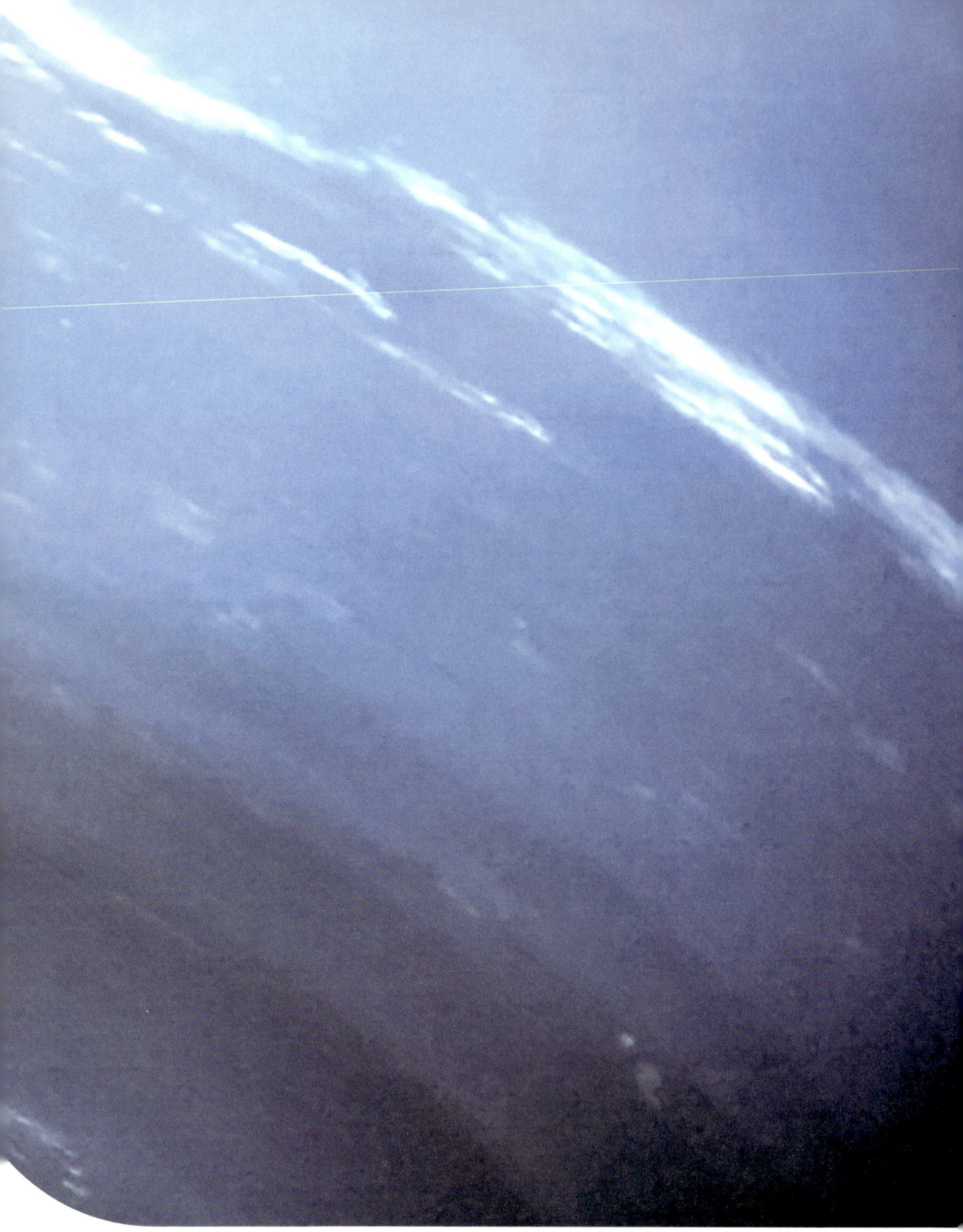

Voyager 2 captured images of storms that appeared as white streaks across Neptune's blue surface.

several days. "Every day, we learned something new," said Ed Stone, lead scientist on the Voyager team.[1] No one had seen Neptune like this before.

Neptune is billions of miles from Earth. It is not visible with the naked eye. It looks like a faint blue dot through telescopes. Before *Voyager 2*, astronomers knew some basic facts about Neptune. But they could see very few details from Earth. They had to send a spacecraft for a closer look.

Voyager 2 had traveled for 12 years before reaching Neptune. It observed the planet for 5 months as it flew past. It saw clouds and giant storms in Neptune's **atmosphere**. It took the first images of Neptune's rings. It also revealed six new moons. *Voyager 2* even took close-ups of Neptune's largest moon, Triton. Before 1989, scientists knew almost nothing about Neptune. Now they had a flood of information about the distant planet.

A GIANT FAMILY

Like Jupiter, Saturn, and Uranus, Neptune is a giant planet. These four planets are found far from the Sun. They are made mostly of gases. They are several times the size of Earth, but they spin much faster. These planets have rings made of ice and rock fragments. And all four have many moons.

Neptune is a planet of extremes. It is the farthest planet from the Sun. It is the smallest and densest giant planet. And it

The giant planet Jupiter is the most massive planet in the solar system.

has the fastest winds in the solar system. Neptune's unique features have caught the attention of many scientists.

CHAPTER ONE

MEET NEPTUNE

Neptune's diameter from pole to pole is 30,250 miles (48,682 km). It is nearly four times the diameter of Earth. Neptune's mass is seventeen times Earth's mass. But Neptune is much less dense than Earth. It spins around its axis of rotation once every 16 hours. Because of its low density and fast spin, Neptune bulges at its equator. It is not a perfect sphere.

Neptune bulges by about 263 miles (423 km) at its equator.

Pluto was once considered the ninth planet of the solar system. But in 2006, most scientists agreed that Pluto did not meet the definition of planet.

On average, Neptune is 2.8 billion miles (4.5 billion km) from the Sun. That is about 30 **astronomical units (AU)**. This makes Neptune the farthest planet from the Sun. Neptune's year is 60,189 Earth days long, or about 165 Earth years.

Neptune's orbit is nearly circular. Its path around the Sun does not cross the paths of any other planet. But it does cross the **elliptical** orbit of the dwarf planet Pluto. This means that sometimes Pluto is closer to the Sun than Neptune is. This was the case from 1979 to 1999.

ICE GIANT

There are two types of planets in the solar system. The rocky planets are Mercury, Venus, Earth, and Mars. They have solid surfaces. The giant planets are Jupiter, Saturn, Uranus, and Neptune. These planets do not have solid surfaces. Instead, they are made mostly of gas. In Jupiter and Saturn, the gas is hydrogen and helium. These two planets are called gas giants.

Uranus and Neptune are smaller than Jupiter and Saturn. They are farther from the Sun. And they are much colder. "We have very little information about them," says Dr. Heidi Hammel.[2] In Uranus and Neptune, the region around the core is made of water, ammonia, and **methane**. These compounds would normally freeze at Uranus and Neptune's distance from

The Colors of Giant Planets

Giant planet atmospheres are made of hydrogen, helium, and hydrogen compounds. A planet's temperature determines which compounds are present. Jupiter and Saturn's atmospheres contain ammonia and water. This creates yellows, reds, whites, and browns. Uranus and Neptune are colder. Their atmospheres contain methane, which appears blue-green.

the Sun. That is why Uranus and Neptune are called ice giants. But the ices in these planets are actually not frozen. Strong gravity compresses them into a hot, dense liquid.

Uranus is larger in diameter than Neptune but lighter in mass. This is because the material making up Uranus is less tightly packed than the material that makes up Neptune.

Neptune's **mantle** is an extreme place. Its upper regions are hot and dense. They have a temperature of about 3,140°F (1,727°C). And they produce many times the pressure as Earth's atmosphere. The core is even hotter and denser.

Laboratory scientists have reproduced diamond rain that resembles the diamond rain that may take place on Neptune.

Temperatures reach about 12,140°F (6,727°C). The pressure is 6 million times Earth's atmospheric pressure.

In Neptune's conditions, scientists think diamond rain may occur. At high temperatures, methane molecules break up into carbon and hydrogen atoms. The carbon atoms join together into crystals like diamonds. These crystals are denser than their surroundings. They fall down toward the core. They enter the hotter, denser region of Neptune. Then they **vaporize** and rise back up. The cycle of falling diamonds and rising gas repeats.

THE WEATHER ON NEPTUNE

Neptune's axis is tilted by 28 degrees. This is very similar to Earth's tilt of 23.5 degrees.

A tilted axis causes parts of a planet to get more sunlight than others. This causes seasons. The average temperature on Neptune is –330°F (–201°C). But Neptune's south pole can be 50°F (10°C) warmer in summer.

Neptune does not have a solid surface. It has no oceans or landmasses. But it is not a simple ball of gas. Storms and clouds race around the planet. They fade as new ones appear.

Neptune's winds are the fastest in the solar system. They can reach more than 1,200 miles per hour (1,900 kmh). The fastest winds on Earth are only 250 miles per hour (400 kmh). "The extreme atmospheric conditions . . . found in the ice giants provide a unique opportunity to

Strong storms on Neptune sometimes appear as dark spots in the atmosphere.

better understand the physics of weather and climate," says planetary scientist Dr. Michael Roman.[3]

RINGS AND MOONS

Neptune's five main rings are made of small pieces of ice and rock. Neptune's rings

NEPTUNE'S RINGS

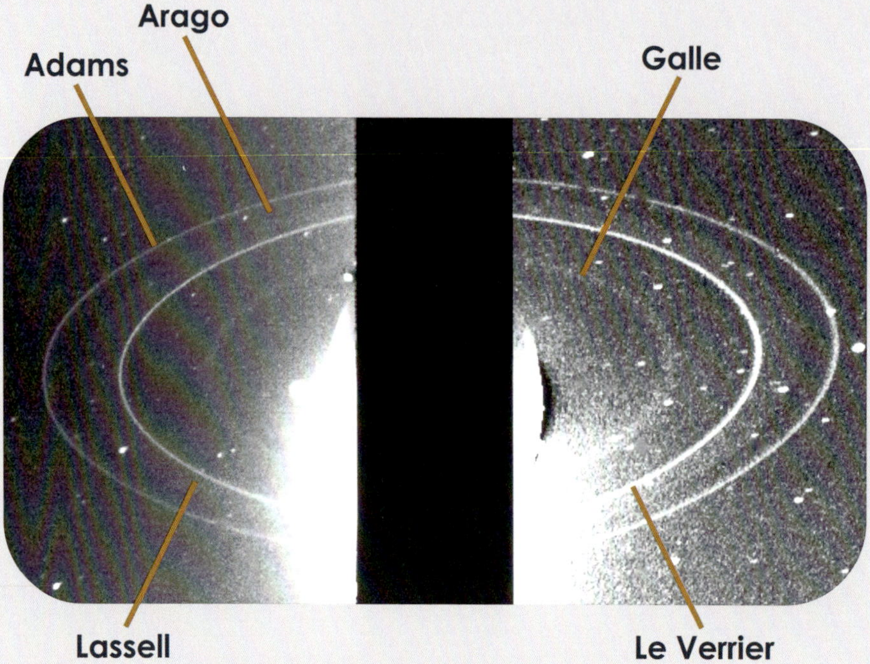

This diagram of Neptune's rings uses images taken by *Voyager 2*. The probe's cameras did not capture Neptune because its brightness would have blocked the rings.

are not as visible as Saturn's. They are thin and dark. The rings are named after early scientists who studied Neptune, such as French astronomer Urbain Le Verrier.

The outer main ring contains four bright arcs. These arcs are places where the material forming the ring clumps together. Some scientists believe these arcs formed after a small moon broke into many pieces. They think this may have happened only a few thousand years ago. Normally, clumps like these should spread out over time. But scientists think the gravity of one of Neptune's moons is holding the clumps together.

Neptune has sixteen moons. The largest, Triton, is nearly the size of Earth's Moon. The others are less than one-fifth Triton's size. The largest is 137 miles (220 km) across. The smallest is 20 miles (34 km) across. Many of Neptune's smaller moons are not spheres. They look like discs or lumpy potatoes. Their gravity is not strong enough to make them round.

CHAPTER TWO

DISCOVERING NEPTUNE

Many of the planets in the solar system can be seen from Earth with the naked eye. These include Mercury, Venus, Mars, Jupiter, and Saturn. Humans have observed them for thousands of years. Watching their movements helped scientists determine the structure of the solar system. But the discovery of Neptune was unique. It was the first planet to be discovered using math.

The invention of the telescope changed astronomy forever. This telescope was used by English scientist Isaac Newton.

This illustration depicts the telescope that William Herschel used to discover Uranus.

In 1781, William Herschel discovered the solar system's seventh planet, Uranus. Uranus is barely visible with the naked eye. It moves slowly in the sky compared to other planets. Herschel discovered Uranus when he was observing stars with a telescope. He saw a tiny speck of light that was not on his maps of the stars. It changed position from one night to the next.

Astronomers observed Uranus for the next 64 years. By then, Uranus had traveled through most of its orbit. But the orbit did not match what scientists predicted. They suspected some other planet's gravity was pulling on Uranus. It was altering Uranus's path.

In 1845, two astronomers independently calculated the new planet's orbit. They were John Couch Adams and Urbain Le Verrier. In 1846, Le Verrier wrote to German astronomer Johann Gottfried Galle. He asked Galle to look for this mysterious planet using a telescope. On the night of

September 23, Galle found it. It was very close to Le Verrier's predicted position. He wrote back to Le Verrier right away. He said, "The planet whose position you indicated really exists."[4]

NAMING NEPTUNE

Several people had played a role in discovering this new planet. They all had ideas for its name. Galle suggested Janus. He is the two-faced Roman god of doorways. English astronomers wanted to name it after Oceanus. He is the Greek god of an imaginary river surrounding Earth.

Le Verrier suggested naming the planet Neptune after the Roman god of the sea. He later changed his mind. He wanted to

Le Verrier (right) discusses astronomy with King Louis Philippe I of France (left).

name it after himself instead. But Neptune won out eventually. It made the most sense. The planet was blue-greenish like the sea. All planets in the solar system besides Earth are named after Roman or Greek gods. Neptune's moons are named after minor ocean gods, goddesses, and sea spirits.

NEPTUNE'S FAMILY

Astronomers were excited about the solar system's eighth planet. Everyone wanted to observe it. But there were difficulties in doing so. Neptune is quite far away from the Sun. The planet is not very bright when viewed from Earth. Earth's atmosphere causes other issues. This layer of gases blurs distant images through telescopes.

Seeing Neptune First

Johann Gottfried Galle was the first to know he was looking at Neptune through a telescope. But people before him saw the planet through telescopes too. Italian astronomer Galileo recorded an observation of Neptune in 1613. But he did not know he was looking at a new planet. He thought it was a star.

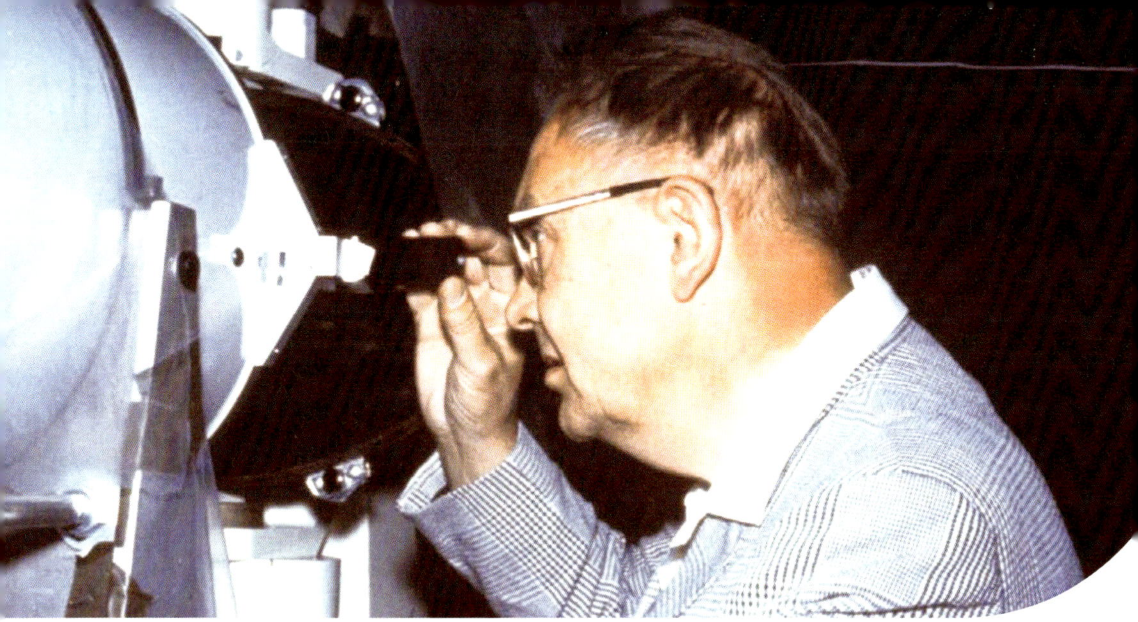

An asteroid belt that lies beyond the orbit of Neptune is named after Gerard Kuiper.

Despite these challenges, William Lassell discovered a moon orbiting Neptune. He discovered it only 17 days after Neptune's discovery. But it would be a long time before more moons were discovered. In 1949, Gerard Kuiper discovered a second moon. They needed names. The first moon was named Triton. He is the son of the Greek god of the sea. The second moon was named Nereid. The nereids are sea spirits in Greek mythology.

CHAPTER THREE

EXPLORING THE BLUE GIANT

In the early 1970s, NASA sent spacecraft to Jupiter and Saturn. *Pioneer 10* and *Pioneer 11* took close-up photos of both planets. They made many measurements. These missions convinced NASA to send the much larger *Voyager* probes to all four giant planets.

Launched in 1977, *Voyager 1* and *2* were identical spacecraft. They carried cameras and scientific instruments that could send

***Pioneer 10* was the first spacecraft to fly by a giant planet. It passed by Jupiter in December 1973.**

data back to Earth. The *Voyager* spacecraft were able to visit the giant planets because all were on the same side of the Sun. This only happens every 175 years.

Voyager 1 and *Voyager 2* visited Jupiter in 1979. *Voyager 1* visited Saturn in 1980. *Voyager 2* reached the planet in 1981. The probes did not have a lot of time

Voyager 2 **captured this image of Saturn and some of its moons as the probe flew by the planet.**

to observe the planets. This is because the probes did not go into a stable orbit around the planets. Instead, they flew close to the planets. The planets' strong gravity then flung the probes in a new direction. *Voyager 1* and *2* took lots of pictures of the planets while they could. They also collected lots of data.

After *Voyager 1*'s visit to Saturn, NASA scientists sent it to one of Saturn's moons. It then continued on a path away from the other planets. The scientists sent *Voyager 2* to Uranus and Neptune next. It used Saturn's gravity to change its course. The spacecraft headed for the more distant planets.

Voyager 2 arrived at Neptune in 1989. At that time astronomers knew Neptune's orbit, color, mass, and size. They knew that Neptune had two moons. Almost everything else was a mystery. The probe observed Neptune from June to October 1989. Most data and pictures were collected from August 21 to 29. The closest approach brought *Voyager 2* within 2,980 miles (4,800 km) of Neptune on August 25.

Scientific Instruments on *Voyager 2*

Voyager 2 studied Neptune with many scientific instruments. These included cameras. NASA also included devices that could detect heat. Some instruments could detect waves of energy other than light, such as radio waves. NASA turned off some of these instruments after *Voyager 2* visited Neptune. They did this to save power.

Voyager 2 remains the only spacecraft ever to visit Neptune. It is also the only spacecraft that has observed all four giant planets. It has provided important insights about these planets. "We didn't know what our solar system looked like until *Voyager* [2] passed Jupiter, Saturn, Uranus and Neptune," says planetary scientist Dr. Carolyn Porco.[5]

DARK AND DISTANT

On Neptune, the Sun's light is 900 times dimmer than on Earth. *Voyager 2* had to work in very low light. It took pictures with long exposures. This meant that it took in more light for each picture. As it sped by Neptune, *Voyager 2* had to rotate to keep its camera aimed in the right direction.

Otherwise, Neptune and its moons would be blurry streaks in the pictures.

Voyager 2 faced other significant challenges. Signals took 4 hours to get from Earth to Neptune. Scientists could not guide *Voyager 2* in real time. NASA designed the probe to collect data automatically. Scientists predicted where *Voyager 2* was going to be at certain times in the future. Then they sent instructions to the probe about what to do when those times came. For example, scientists knew when *Voyager 2* would be close to Neptune. They told the probe where to point the camera far in advance.

By the time *Voyager 2*'s signal reached Earth, it was very weak. NASA needed to improve its technology on Earth to catch

Scientists continue to build new dishes to expand the power of the Deep Space Network.

these signals. While the probe traveled from Uranus to Neptune, NASA expanded the Deep Space Network. This is a group of satellite dishes on Earth. NASA added more dishes and made some larger. These dishes made getting information from *Voyager 2* easier.

Another challenge *Voyager 2* faced was its age. It had been built to last only 5 years. In 1989, it was 12 years old.

Its instruments did not work perfectly. It did not communicate well. And its generator was running out of energy. In spite of these challenges, *Voyager 2* accomplished its mission.

SURPRISE DISCOVERIES

As *Voyager 2* approached Neptune, it sent pictures to Earth. The pictures were the best images of Neptune anyone had ever seen. The first surprise discovery was a new moon, Proteus. Three more moons soon followed. The probe discovered six moons in all. Some were orbiting in the planet's rings. *Voyager 2* discovered five rings around Neptune.

Voyager 2 images revealed Neptune's Great Dark Spot. This was a hurricane-like storm in Neptune's atmosphere. It was larger than Earth. *Voyager 2* discovered two other storms, the Small Dark Spot and Scooter. Scientists named Scooter after the way it scooted around Neptune faster than the Great Dark Spot. The probe took pictures of clouds in Neptune's

***Voyager 2* captured this image of Neptune's Small Dark Spot.**

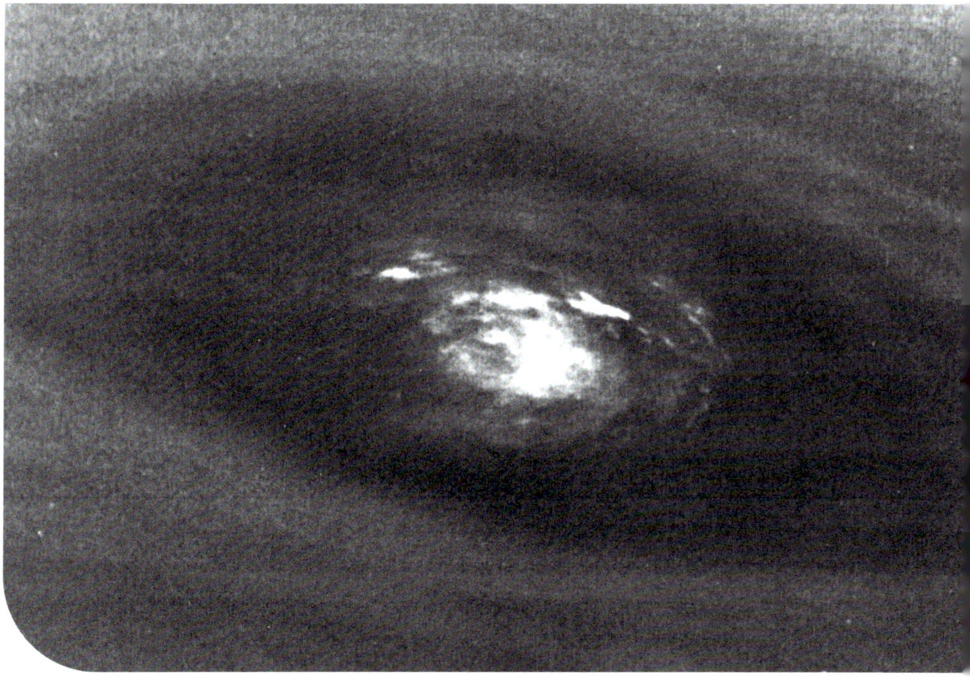

upper atmosphere. It also measured wind speeds and the planet's **magnetic field**.

The *Voyager 2* scientists wanted close-up pictures of Triton. They changed the spacecraft's path so that it would pass within 25,000 miles (40,000 km) of the moon. It took detailed images of parts of Triton. They revealed rugged regions of Triton's surface. *Voyager 2* even watched a volcano erupt.

On its way out of Neptune's neighborhood, *Voyager 2* took more pictures. It observed sunlight through the atmospheres of Neptune and Triton. In total, *Voyager 2* took more than 9,000 images of the Neptune system.

Since *Voyager 2*'s historic visit, two powerful telescopes have been launched into space. The Hubble Space Telescope (HST) orbits Earth. The James Webb Space Telescope (JWST) orbits the Sun. Earth's atmosphere blurs images and blocks certain kinds of light. This means space telescopes can see better than telescopes

The Hubble Space Telescope is about 43.5 feet (13.2 m) long.

To work properly, the James Webb Space Telescope must be kept at a very low temperature. The telescope is designed to stay at about –390°F (–235°C).

on Earth. HST images of Neptune in 1994 showed that the Great Dark Spot had disappeared. Later images showed spots in constantly changing patterns.

JWST observes Neptune in **infrared** light. The human eye cannot see this light. The telescope captures infrared light to see deep into Neptune's atmosphere.

JWST can also see Neptune's rings and moons. "Right now, the only way to study ice giants is with the biggest and best telescopes in the world," says Dr. Hammel.[6] Until another spacecraft is sent to Neptune, JWST will provide the most detailed views of the planet.

FUTURE MISSIONS

After its visit to Neptune, *Voyager 2* kept flying. In 2018, it left the solar system. It does not have much power left, but it was still sending data to Earth in 2024.

NASA has proposed two missions to send spacecraft to Neptune and its moons. They are *Trident* and *Neptune Odyssey*. The Chinese National Space Administration is also planning a Neptune mission.

NEPTUNE'S MOONS

Astronomers have discovered sixteen moons of Neptune. About half formed along with the planet and its rings. The rest formed elsewhere. They came too close and were captured by Neptune's gravity.

Triton is Neptune's largest moon. It has a diameter of 1,680 miles (2,700 km). That's about three-quarters the size of Earth's Moon. Triton has a core of rock and metal. Its mantle is frozen water. Triton is one of

This NASA illustration shows what Neptune might look like from the surface of Triton.

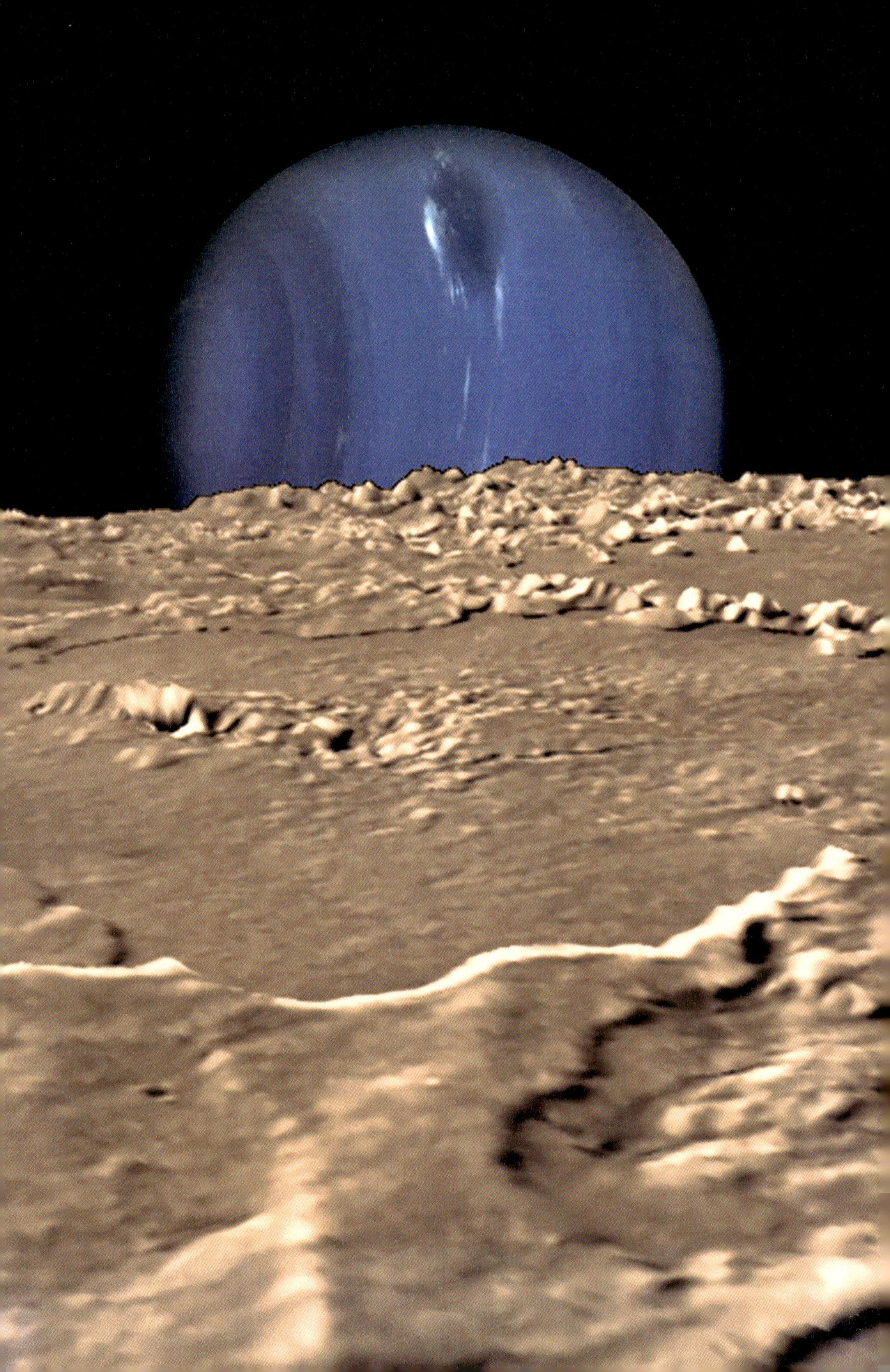

the coldest places in the solar system. Its average temperature is –391°F (–235°C). Triton's surface has craters and flowing lava.

Voyager 2 **captured this detailed image of Triton's surface in 1989.**

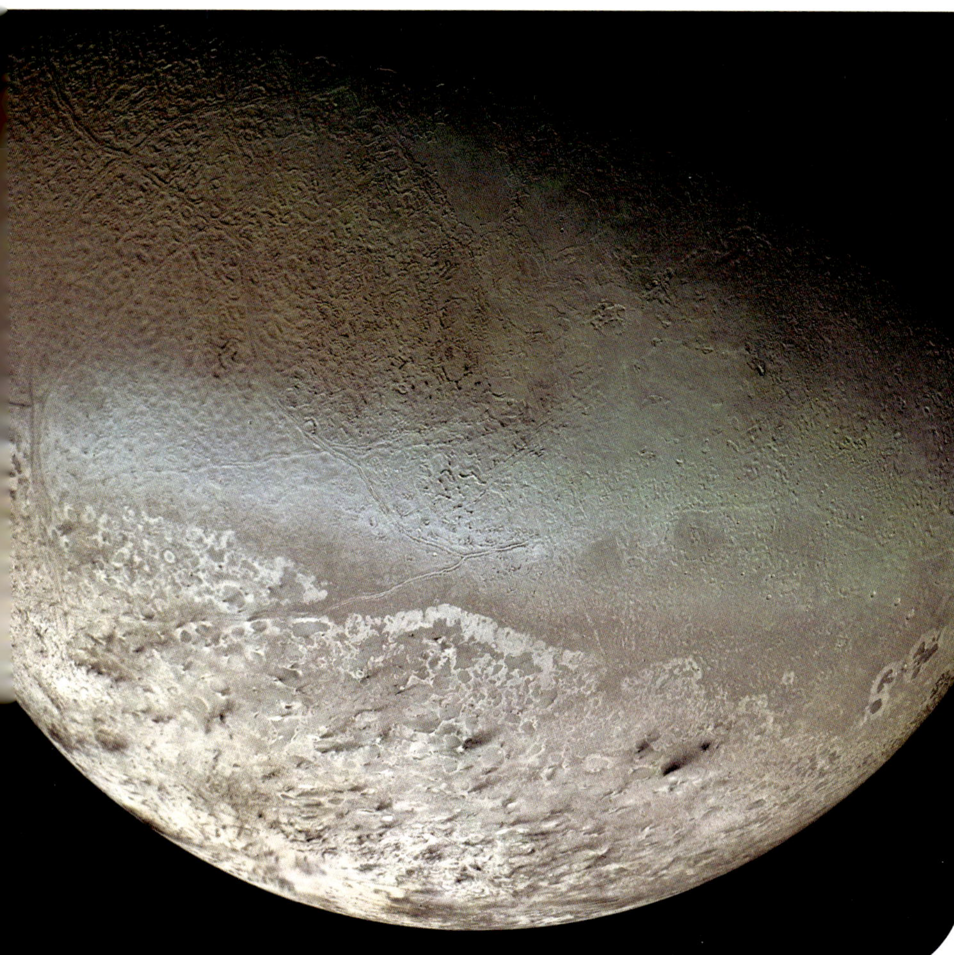

It is one of the few volcanically active objects in the solar system.

Triton's atmosphere is composed of nitrogen and methane. The atmosphere comes from ice volcanoes. They spew these gases from Triton's interior. The gases freeze instantly. Then they snow down onto the surface. As a result, Triton is covered by a crust of bright nitrogen frost.

Most large moons in the solar system spin and orbit in the same direction. They formed along with their parent planets from the spinning cloud of gas and dust that created the Sun. But Triton orbits in the opposite direction from Neptune's spin. This is called a retrograde orbit. Its orbit makes a steep angle with Neptune's equator. Its nearly circular orbit keeps the moon about

Asteroids are space rocks left over from the solar system's formation. Asteroid impacts are a common cause of craters on other space objects.

220,438 miles (354,760 km) from Neptune. It takes Triton about 6 days to complete one backward loop around the planet.

Triton is very similar to Pluto. This fact and Triton's unusual orbit support the theory that Triton is a captured object. At one time, Triton was probably a dwarf planet.

NEREID

Nereid was the second moon of Neptune to be discovered. It is about 105 miles (170 km) across. That makes it the third-largest moon of Neptune. Like all of

Voyager 2 took this image of Nereid as it flew by Neptune and its moons.

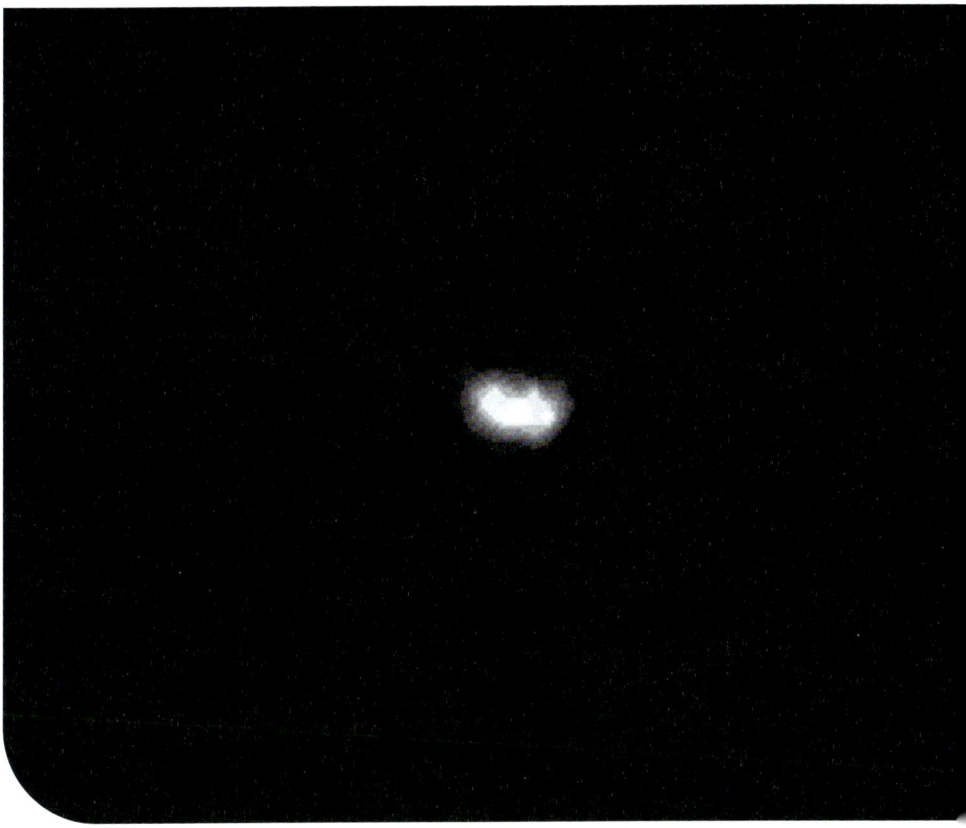

Neptune's moons, it is most likely made of ice and rock.

Like Triton, Nereid has an unusual orbit. It is a very elongated ellipse. At its closest it comes within about 841,100 miles (1,353,600 km) of Neptune. At its most distant it is nearly 6 million miles (9.6 million km) away. It may once have had a more circular orbit. The capture of Triton probably disturbed Nereid's orbit.

REGULAR MOONS

Neptune's remaining moons can be divided into two groups. There are seven regular moons and seven irregular moons. The regular moons formed out of the same gas that created Neptune. They are close to Neptune and have fairly circular orbits.

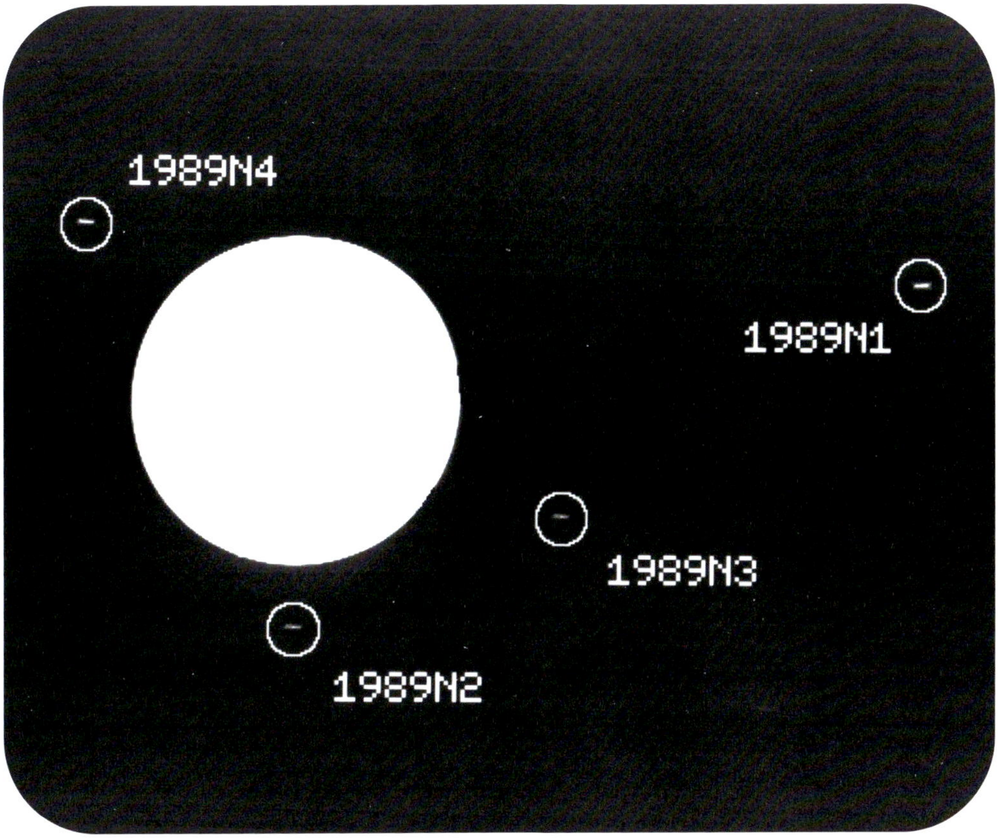

This labeled *Voyager 2* image shows Proteus, Larissa, Despina, and Galatea before they were officially named.

Most were discovered by *Voyager 2*. These moons are not spherical. They are covered in craters.

Naiad is sometimes called the lost moon. After its discovery in 1989, it was not spotted again until 2013. It orbits only 30,000 miles (48,000 km) from Neptune.

That makes it difficult to observe Naiad from Earth.

Thalassa is another regular moon. It is shaped like a disc. Naiad and Thalassa's orbits are shrinking. The moons will eventually fall into Neptune.

Despina and Galatea orbit within Neptune's ring system. Their gravity keeps the rings separate from each other. Galatea may be causing the outermost ring to separate into four arcs.

Larissa was briefly spotted from Earth in 1981. *Voyager 2* confirmed its discovery. Hippocamp is only 20 miles (34 km) across. It was found in HST images.

Proteus is larger than Nereid. But it is much darker and closer to Neptune. This may be why it was not discovered until

Voyager 2 arrived. At about 73,000 miles (118,000 km) from Neptune, it is the most distant regular moon.

IRREGULAR MOONS

The irregular moons are much smaller than the regular moons. They have tilted, oval-shaped orbits. They orbit hundreds of times farther from Neptune than the

The Kuiper Belt

The Kuiper Belt is a large disc of chunks of ice and rock. It extends from Neptune's orbit to about 50 AU from the Sun. The objects in the Kuiper Belt are left over from the solar system's birth. Some of Neptune's moons may have been captured objects from this region. These include Triton.

This illustration depicts an asteroid belt. In reality, the rocks within the Kuiper Belt are far more spread apart than this image suggests.

regular moons. All were discovered by powerful Earth-based telescopes and the HST. The JWST continues to track them.

Astronomers believe that the irregular moons are captured objects. The closest to Neptune is Halimede. It orbits about

10,365,000 miles (16,681,000 km) away. Halimede has a retrograde orbit.

The orbits of Neso, Psamathe, and an unnamed moon are very similar. They are the most distant irregular moons. They orbit up to 31 million miles (50 million km) away from Neptune. They may once have been a single moon. This moon may have broken apart when it collided with comets or asteroids. Like Triton and Halimede, they have retrograde orbits.

Sao, Laomedeia, and another unnamed moon may also be pieces of a single moon. These broken moons show that the early solar system was a chaotic place. “Neptune seems to hold more than its share of information about the formation of our own solar system,” says Dr. Hammel.[7]

GLOSSARY

astronomers

scientists who study space

astronomical units (AU)

units representing the average distance between Earth and the Sun, equal to about 93 million miles (150 million km)

atmosphere

the gases surrounding a planet

elliptical

shaped like an ellipse

infrared

a kind of light that cannot be seen by humans but can be measured using devices

magnetic field

the area around a magnetic material within which the force of magnetism works

mantle

the region of a planet between its core and its crust

methane

a chemical made of carbon and hydrogen

vaporize

to go from a solid or liquid state to a gaseous state

SOURCE NOTES

INTRODUCTION: THE END OF A LONG VOYAGE

1. Quoted in Anthony Greicius, "30 Years Ago: *Voyager 2*'s Historic Neptune Flyby," *NASA*, August 22, 2019. www.nasa.gov.

CHAPTER ONE: MEET NEPTUNE

2. Quoted in "Icy Giants Uranus and Neptune as Observed from JWST, ft. Dr. Heidi B. Hammel," *YouTube*, uploaded by SETI Institute, June 13, 2023. www.youtube.com.

CHAPTER TWO: DISCOVERING NEPTUNE

3. Quoted in Charles Q. Choi, "Neptune: A Guide to the Windy Eighth Planet from the Sun," *Space.com*, April 17, 2023. www.space.com.

CHAPTER THREE: EXPLORING THE BLUE GIANT

4. Quoted in Alan Boyle, "One 'Year' Ago, Neptune Was Found," *NBC News*, July 12, 2011. www.nbcnews.com.

5. Quoted in "A Spirited Conversation with Carolyn Porco," *The Planetary Society*, November 13, 2019. www.planetary.org.

6. Quoted in "Icy Giants Uranus and Neptune as Observed from JWST, ft. Dr. Heidi B. Hammel."

CHAPTER FOUR: NEPTUNE'S MOONS

7. Quoted in Curtis Rist, "Neptune Rising," *Discover*, September 1, 2000. www.discovermagazine.com.

FOR FURTHER RESEARCH

BOOKS

Dr. Bruce Betts, *Solar System Reference for Teens*. Callisto, 2022.

KS Mitchell, *The Gas Giants: Jupiter, Saturn, Uranus, and Neptune*. BrightPoint Press, 2023.

Heather C. Morris, *Uranus*. BrightPoint Press, 2026.

INTERNET SOURCES

"Neptune," *National Air and Space Museum*, n.d. https://airandspace.si.edu.

"New Webb Image Captures Clearest View of Neptune's Rings in Decades," *NASA*, September 21, 2022. www.nasa.gov.

"Triton, Neptune's Largest Moon," *The Planetary Society*, n.d. www.planetary.org.

WEBSITES

NASA: Neptune

https://science.nasa.gov/neptune

This NASA website provides general information about Neptune. It also discusses the history of its exploration.

The Schools' Observatory

www.schoolsobservatory.org

The Schools' Observatory provides educational resources related to space. It also provides visitors with career paths in space science.

Solar System Scope

www.solarsystemscope.com

Solar System Scope is an online simulation of the solar system. Clicking on objects such as Neptune provides visitors with information about the object.

INDEX

IMAGE CREDITS

Cover: © NASA
5: © NASA
7: © NASA
8: © NASA
11: © NASA
13: © NASA
14: © NASA
17: © NASA
18: © Peter Simoncik/Shutterstock Images
21: © NASA
22: © NASA
25: © Science Source
26: © Photo Researchers/Science History Images/Alamy
29: © Chronicle/Alamy
31: © NASA
33: © NASA
34: © NASA
39: © NASA
41: © NASA
43: © NASA
44: © NASA
47: © NASA
48: © NASA
50: © NASA
51: © NASA
53: © NASA
56: © Anterovium/Shutterstock Images

ABOUT THE AUTHOR

Meg Thacher teaches astronomy at Smith College in Massachusetts. She is the author of *Sky Gazing: A Guide to the Moon, Sun, Planets, Stars, Eclipses, and Constellations.* She writes books and articles about science, classical music, and mythology—but mostly astronomy. She arrived at astronomy graduate school around the same time that *Voyager 2* arrived at Neptune.